MAR 2019

WE LOVE Squares!

BY BEATRICE HARRIS

Gareth Stevens
PUBLISHING

first concepts

The cracker is
a square.

3

The gift is a square.

5

The pillow is
a square.

7

The plate is a square.

9

The candy is a square.

11

The clock is a square.

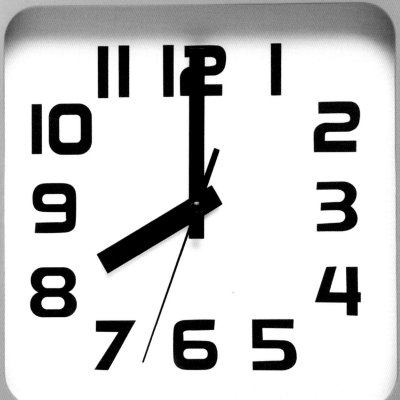

13

The paper is a square.

15

The table is a square.

17

The picture is
a square.

19

The stones
are squares.

Point to the square.
We love squares!

23

Please visit our website, www.garethstevens.com. For a free color catalog of all our high-quality books, call toll free 1-800-542-2595 or fax 1-877-542-2596.

Cataloging-in-Publication Data

Names: Harris, Beatrice.
Title: We love squares! / Beatrice Harris.
Description: New York : Gareth Stevens Publishing, 2018. | Series: Our favorite shapes | Includes index.
Identifiers: ISBN 9781538210017 (pbk.) | ISBN 9781538210031 (library bound) | ISBN 9781538210024 (6 pack)
Subjects: LCSH: Square–Juvenile literature. | Shapes–Juvenile literature.
Classification: LCC QA482.H37 2018 | DDC 516'.154–dc23

Published in 2018 by
Gareth Stevens Publishing
111 East 14th Street, Suite 349
New York, NY 10003

Copyright © 2018 Gareth Stevens Publishing

Designer: Bethany Perl
Editor: Therese Shea

Photo credits: Cover, p. 1 (graph paper) Roobcio/Shutterstock.com; p. 1 (window) N-studio/Shutterstock.com; p. 1 (gingham fabric) daizuoxin/Shutterstock.com; p. 1 (tile) patinya mungmai/Shutterstock.com; p. 3 mayakova/Shutterstock.com; p. 5 StudioByTheSea/Shutterstock.com; p. 7 Africa Studio/Shutterstock.com; p. 9 Kobkob/Shutterstock.com; p. 11 Irina Rogova/Shutterstock.com; p. 13 5 second Studio/Shutterstock.com; p. 15 Skylines/Shutterstock.com; p. 17 (cat) Angela Waye/Shutterstock.com; p. 17 (frame) IS MODE/Shutterstock.com; p. 19 Hirunya/Shutterstock.com; p. 21 CK2 Connect Studio/Shutterstock.com; p. 23 Ilona Belous/Shutterstock.com.

Printed in China

CPSIA compliance information: Batch #CW18GS: For further information contact Gareth Stevens, New York, New York at 1-800-542-2595.